THAT'S ANCIENT!

The Discoveries of
ANCIENT CHINA

BY JANEY LEVY

Gareth Stevens
PUBLISHING

Library of Congress Cataloging-in-Publication Data

Names: Levy, Janey, author.
Title: The discoveries of ancient China / Janey Levy.
Description: New York : Gareth Stevens Publishing, [2022] | Series: That's ancient! | Includes bibliographical references and index.
Identifiers: LCCN 2020045706 (print) | LCCN 2020045707 (ebook) | ISBN 9781538265673 (library binding) | ISBN 9781538265659 (paperback) | ISBN 9781538265666 (set) | ISBN 9781538265680 (ebook)
Subjects: LCSH: Inventions—China—History—Juvenile literature. | Technology—China—History—Juvenile literature. | Civilization—Chinese influences—Juvenile literature.
Classification: LCC T27.C5 L46 2022 (print) | LCC T27.C5 (ebook) | DDC 609.3—dc23
LC record available at https://lccn.loc.gov/2020045706
LC ebook record available at https://lccn.loc.gov/2020045707

First Edition

Published in 2022 by
Gareth Stevens Publishing
29 E. 21st Street
New York, NY 10010

Copyright © 2022 Gareth Stevens Publishing

Designer: Katelyn E. Reynolds
Editor: Therese Shea

Photo credits: Cover, p. 1 gyn9037/Shutterstock.com; cover, pp. 1–32 (burst) Dawid Lech/Shutterstock.com; cover, pp. 1–32 (clouds) javarman/Shutterstock.com; p. 5 Andrei Minsk/Shutterstock.com; pp. 7 (top), 9, 15 (left) China Photos/Getty Images; p. 7 (bottom) drs2biz (https://www.flickr.com/photos/drs2biz/)/PericlesofAthens/Wikipedia.org; p. 8 Nagualdesign/Wikipedia.org; p. 10 Three Lions/Getty Images; p. 11 Zhang Peng/LightRocket via Getty Images; p. 13 (left) Mark Ralston/AFP via Getty Images; p. 13 (right) Sam Yeh/AFP via Getty Images; p. 15 (right) Goh Chai Hin/AFP via Getty Images; p. 16 Garsya/Shutterstock.com; p. 17 J J Osuna Caballero/Shutterstock.com; p. 18 BabelStone/Wikipedia.org; p. 19 Originally uploaded by 屋 in Chinese Wikipedia/MichaelFrey/Wikipedia.org; p. 21 (left) Hulton Archive/Getty Images; p. 21 (right) PericlesofAthens/Wikipedia.org; p. 21 (bottom) Philippe Lopez/AFP via Getty Images; pp. 23, 27 SSPL/Getty Images; p. 24 Marzolino/Shutterstock.com; p. 25 Charlesdrakew/Wikipedia.org; p. 27 (inset) Cmglee/Wikipedia.org; p. 29 (left) This file comes from Wellcome Images (http://wellcomeimages.org/), a website operated by Wellcome Trust, a global charitable foundation based in the United Kingdom/Fæ/Wikipedia.org; p. 29 (right) Guillaume Souvant/AFP via Getty Images.

Printed in the United States of America

Some of the images in this book illustrate individuals who are models. The depictions do not imply actual situations or events.

CPSIA compliance information: Batch #CWGS22: For further information, contact Gareth Stevens, New York, New York, at 1-800-542-2595.

Find us on

CONTENTS

Words in the glossary appear in **bold** type the
first time they are used in the text.

Ancient

CHINA

What comes to mind when you hear people talk about China? Perhaps you imagine a faraway, mysterious travel destination. Or perhaps you think of it more as a nation of great economic and political power. But what do you know about China's rich history?

The ancient world saw the rise of many different **cultures**. Over time, most cultures collapsed and disappeared. Ancient China was different. It gave rise to a culture that still exists today.

You might believe ancient Chinese culture was so different from our modern culture that the two have nothing in common. But you'd be amazed. We use ancient Chinese inventions on a regular basis—inventions such as paper and tea. You'll learn lots more about ancient Chinese inventions and discoveries inside this book.

THAT'S FASCINATING!

Early villages first appeared along the Yellow River in China around 5000 BCE.

THE FIRST DYNASTY

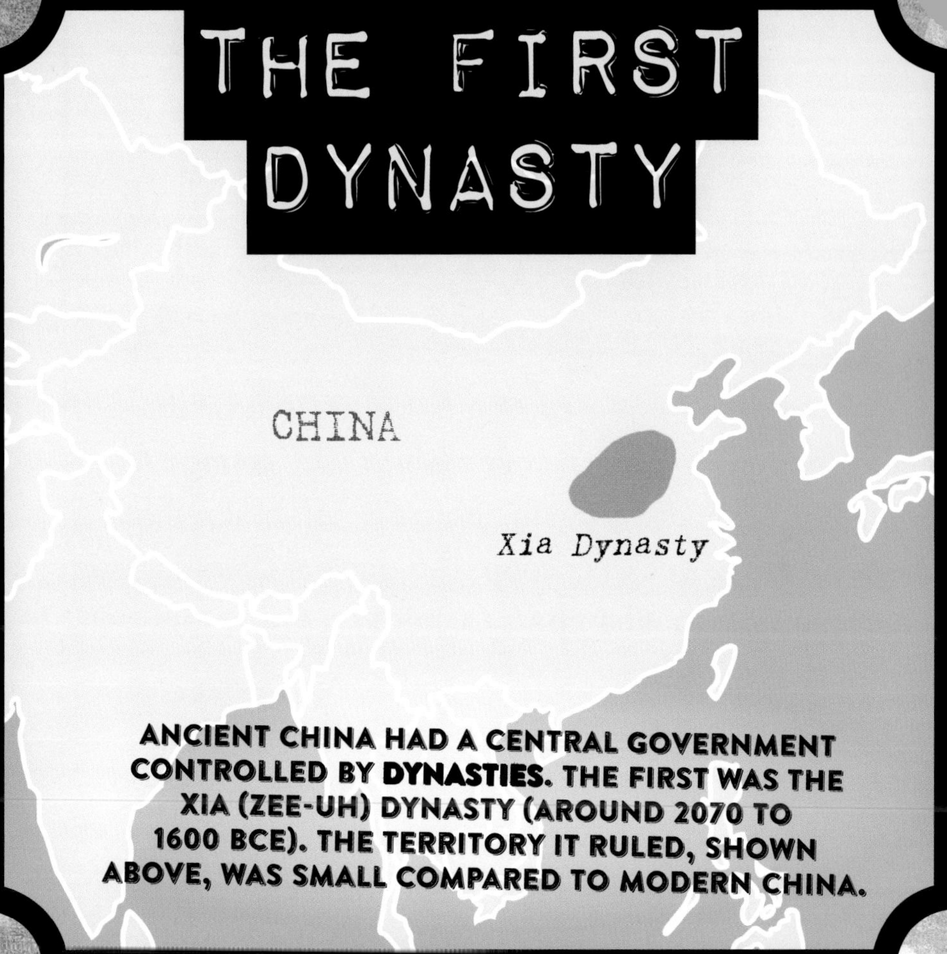

CHINA

Xia Dynasty

ANCIENT CHINA HAD A CENTRAL GOVERNMENT CONTROLLED BY **DYNASTIES.** THE FIRST WAS THE XIA (ZEE-UH) DYNASTY (AROUND 2070 TO 1600 BCE). THE TERRITORY IT RULED, SHOWN ABOVE, WAS SMALL COMPARED TO MODERN CHINA.

The Xia Dynasty

It was long believed that the Xia dynasty was just a myth, created to explain and justify why dynasties had changed throughout Chinese history. According to this interpretation, heaven granted rulers their authority to rule. Xia dynasty rulers created everything needed for civilization. But human failings caused a collapse, and a new dynasty came to power. Discoveries in the 1960s and 1970s challenged this interpretation. Archaeologists found buildings that matched those described by historians writing about the Xia dynasty.

From a Cocoon to

A FABRIC

You know what a cocoon is, right? Caterpillars create them as the place within which they change into a butterfly or moth. In China, there's a caterpillar that creates a very special cocoon. The ancient Chinese used the cocoon's long, fine fibers to create a luxury fabric still valued today—silk.

Archaeologists discovered evidence that the ancient Chinese were raising these caterpillars, called silkworms, by about 3600 BCE. The oldest known examples of woven silk date to about 2700 BCE.

Silk was used to make clothing for the royal family and members of the upper social classes. It was also used for fans, for banners, and as a surface for writers and artists to work on. It became a valued trade good in China and internationally.

THAT'S FASCINATING!

Silk was also used to make kites, which were developed in China around 1000 BCE. Kites were originally used for military purposes, such as sending messages and measuring distances.

Sophisticated Silk Tech

The ancient Chinese didn't just put silkworms in a room, feed them, and leave them to make cocoons. They had an advanced understanding of the silkworms' needs and used that to control the silkworms and improve the silk. They altered the temperature of the silkworms' surroundings to speed up or slow down their growth. They raised different types of silkworms and bred them with each other to raise worms whose silk would have new qualities the weavers wanted.

TEATIME

Lots of people around the world enjoy tea. It's something many people consider a traditional British drink. However, it was the ancient Chinese who discovered tea.

According to legend, tea was discovered accidentally. It happened when tea leaves fell from a tree into Emperor Shennong's pot of boiling water in 2737 BCE. However, actual evidence shows tea drinking began around 1800 BCE.

At first, tea was used as a medicine. Tea leaves were boiled in water along with other ingredients, such as butter, herbs, and spices. Sometimes, tea was made into a paste and applied to skin to relieve pain. It wasn't until the Tang dynasty (618 to 907 CE) that tea came to be widely enjoyed as a drink.

THIS IMAGE SHOWS EMPEROR SHENNONG, THE LEGENDARY DISCOVERER OF TEA.

THAT'S FASCINATING!

The oldest tea tree in the world is in Lincang, China, and is about 3,200 years old!

tea leaves

Tea Bricks

If you drink tea, you know it's made today using dried tea leaves. But that's not how the ancient Chinese made tea. Loose tea leaves take up lots of storage space and are harder to transport.
So the ancient Chinese pressed tea into bricks. To make tea, they ground part of the brick into powder. Then they added the powder to the water in their kettle to boil or put the powder in a cup and added hot water to it.

Save It for a
RAINY DAY

How do you keep dry if you're outside on a rainy day? Maybe you wear a raincoat with a hood to protect yourself from the rain. Or maybe you carry that useful device—the umbrella. Have you ever wondered who invented the umbrella? You can thank the ancient Chinese for helping keep you dry.

Umbrellas were based on parasols, which appeared around 1500 BCE in ancient Egypt. Parasols were designed to give shade from the sun; they weren't intended for protection from rain. The earliest Chinese parasols were made of silk on frames made of bamboo or the bark of mulberry trees.

In the 11th century BCE, the Chinese had an idea to improve parasols. They made them **waterproof** by applying oil to the silk. And that was how umbrellas came to be!

The Chinese are credited with inventing the collapsible umbrella in the 4th century BCE.

AROUND THE 2ND CENTURY BCE, THE CHINESE ALSO BEGAN TO MAKE OILED PAPER UMBRELLAS. LIKE THE MODERN ONES SHOWN HERE, THEY WERE OFTEN DECORATED.

The Umbrella as a Social Symbol

Umbrellas had practical uses, but they had symbolic ones as well. Making umbrellas took a lot of work, so they were scarce and costly. This meant at first only Chinese royalty and nobility carried umbrellas. Later, as more people could afford umbrellas, color became a sign of rank. Only royalty could carry red and yellow umbrellas; everyone else used blue. And a simple umbrella wouldn't do for high-ranking individuals. They used umbrellas with many tiers, or levels!

Do you like Italian **pasta**? Perhaps you enjoy the Japanese noodle soup called ramen (RAH-muhn). Or maybe you love the Polish dumplings called pierogies (puh-ROH-geez). Noodles of some sort can be found all over the world. However, they made their first appearance in ancient China.

The earliest written Chinese accounts of noodles appeared between 500 BCE and 400 BCE. That's around the same time wheat became an established crop there. Why does that matter? Wheat, which contains the protein gluten, is needed for making the flour required for noodles. It was also around this time that the equipment for grinding wheat into flour appeared in China.

The earliest noodles didn't look like what we think of as noodles. They were just little pieces of dough thrown into a pot of boiling water!

THAT'S FASCINATING!

Pasta is often dried and stored before cooking and eating. Chinese noodles are made fresh, to be cooked and eaten immediately.

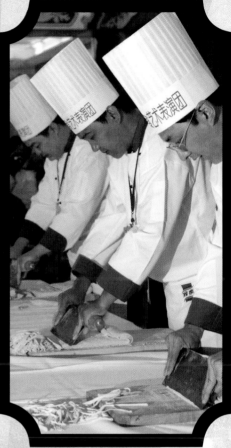

The Noodle-Pasta Question

It seems likely that noodles spread from China to Korea, Japan, and Turkey along the trade routes. So did they make it all the way to Italy as well and become pasta? No one is sure. Sometimes people in different parts of the world have the same idea independently, and that may be what happened here. One expert thinks Italian pasta may have been influenced by **Middle Eastern** noodles, which appeared after the 5th century CE.

Not Just

CHOPSTICKS

Have you ever eaten in a Chinese restaurant? If you have, you know that Chinese people eat with chopsticks. In case you're not familiar with chopsticks, they're slender sticks held between the thumb and fingers of one hand and used to carry food to the mouth. Since Chinese people normally use chopsticks, it might surprise you to learn ancient China was one of the places—along with ancient Egypt and Greece—where the first forks appeared.

Archaeologists uncovered a fork—known as a *cha* in Chinese—at a Xia dynasty (around 2070 to 1600 BCE) site in northwestern China. Forks were also discovered in excavations at Shang dynasty (1600 to 1046 BCE) and Zhou dynasty (1046 to 256 BCE) sites.

These early forks weren't made from metal like modern forks. Rather, they were carved from animal bones or wood.

THAT'S FASCINATING!

Forks didn't reach Europe until the 11th century, and they were regarded with suspicion at the time. They didn't become popular there until the 16th century.

THE TOMB OF THE FIRST EMPEROR OF THE QIN DYNASTY (221–206 BCE) IS FAMOUS FOR ITS THOUSANDS OF LIFE-SIZE CLAY FIGURES OF WARRIORS (BELOW). MOST PEOPLE DON'T KNOW IT ALSO CONTAINED A FORK WITH THE EMPEROR'S PERSONAL SYMBOL ON IT!

chopsticks

Why Chopsticks?

If the ancient Chinese had forks, why did they choose chopsticks? There were several factors. Population growth and limited resources produced a new cooking style. Food was chopped into small pieces to allow quick cooking that conserved fuel. Thus forks—and knives—weren't needed for eating. Chopsticks were practical and effective. Also influential were the teachings of the ancient Chinese philosopher Confucius (551 to 479 BCE). He said sharp tools such as knives and forks represented violence and didn't belong at the dining table.

ARE WE GOING?

Have you ever gone on a camping or hiking trip where you needed to determine what direction you were going? Then you probably used a compass. The oldest, most common type is the magnetic compass. It contains a magnetized needle that rotates freely to line up with Earth's **magnetic field**. The ends point to magnetic north and south.

You can thank the ancient Chinese for the magnetic compass. They invented it between the 2nd century BCE and 1st century CE. But they weren't using it to find the direction they were going. The ancient Chinese used the compass to decide how to bring certain activities into harmony with nature. For example, they might use a compass to find the best place to bury someone or build a house.

THAT'S FASCINATING!

It wasn't until the 11th century that the Chinese began to use compasses for navigation. That's over 1,000 years after they invented the compass!

THIS IS A COMMON ANCIENT CHINESE COMPASS. THE SPOON IS THE "NEEDLE," AND THE HANDLE ALWAYS POINTS SOUTH. THESE COMPASSES WERE CALLED SOUTH-POINTERS.

Making Magnetic Needles

The earliest Chinese compasses used lodestone, a naturally occurring magnetic iron ore, for the spoons in the south-pointers. By the 7th or 8th century CE, the Chinese had found ways to create their own magnetic needles. They had learned how to magnetize iron needles by rubbing them against magnetized iron ore. They had also discovered they could create magnetic needles by heating and then cooling needles and then lining them up with Earth's magnetic field.

Something to
WRITE ON

You probably use paper every day. You either write on it or print out something from a computer. It's such a common item that most people hardly think about it. But have you ever wondered who invented paper? You can thank the ancient Chinese.

The very earliest, primitive paper was made in the 2nd century BCE. It was probably invented accidentally. Clothes made of **hemp** fiber were left in the wash water too long, leaving behind a residue, or debris. That residue was pressed to force out the water, creating paper.

In 105 CE, Cai Lun, director of the Imperial Workshops at Luoyang, improved paper. He made a pulp, or mush, by soaking mulberry tree bark. He then pressed the pulp into sheets, which were then dried on wooden frames or screens.

THAT'S FASCINATING!

Cai Lun later made his paper even better by adding hemp and old fish nets to the mulberry tree bark.

THESE ARE THE FIVE STEPS FOR MAKING PAPER DESCRIBED BY CAI LUN.

More Chinese Paper Creations

Writing paper isn't the only paper invention to thank the Chinese for. They invented toilet paper. Its first recorded use was in 851 CE. Toilet paper didn't get "invented" in the Western world until 1857. The Chinese invented paper money during the Tang dynasty (618 to 907 CE). Before that, money was in the form of coins. And restaurant menus, printed on paper, were created as early as the Song dynasty (960 to 1279).

Explosive
ENTERTAINMENT

What would the Fourth of July be without fireworks? It's hard to imagine, isn't it? People like to use fireworks to celebrate many occasions. But as much as modern people enjoy and use fireworks, they were invented long ago—by the ancient Chinese.

The first fireworks, which were created as early as 200 BCE, were nothing like modern fireworks. Green bamboo caused explosions when it burned in fire. The bamboo had air pockets inside it that expanded in the heat and finally burst the bamboo open with a loud BANG. This was thought to be useful because it scared off evil spirits. Fireworks that more resembled modern displays appeared during the Song dynasty (960 to 1279).

Li Tian, a Chinese monk who lived in the Tang dynasty (618 to 907), invented firecrackers by filling paper tubes with **gunpowder**. Today, he's remembered as the father of firecrackers.

THAT'S FASCINATING!

Gunpowder was discovered during the Tang dynasty. Religious philosophers called Daoists accidentally created it when they tried to develop a drug that would let the emperor live forever.

Fireworks to Rockets

You know astronauts are launched into space using a rocket, right? Well, the ancient Chinese invented rockets—using their fireworks. For a device to qualify as a rocket, it must be **self-propelled**. The earliest description of such a device appeared in 1264. It was a tube-shaped firework, probably of bamboo, filled with gunpowder. When it was lit, the gases escaped through a hole and propelled it around the ground in all directions. And that gave it its name, the ground rat.

The Earth Is

SHAKING!

Earthquakes can be scary, deadly, and very destructive. Unfortunately, it's impossible to forecast them. But scientists constantly study earthquakes to improve their understanding of them. An important tool for studying earthquakes is the seismograph (SYZ-muh-graf). And a scholar named Zhang Heng invented an early version in China in 132 CE.

The ancient Chinese seismograph resembled a large bronze vase, with eight dragon heads sticking out around its upper part. The dragons pointed in the eight principal directions: north, south, east, west, southeast, northeast, southwest, and northwest. Each dragon held a ball in its mouth, and sitting on the ground below each dragon was a frog with an open mouth. When an earthquake occurred, the dragon's head pointing in that direction opened, and the ball dropped into the frog's mouth, making a noise.

THAT'S FASCINATING!

Zhang Heng's "dragon jar" wasn't the kind of vase you could sit on a shelf. According to historical accounts, it was 6 feet (1.8 m) across!

How Exactly Did the "Dragon Jar" Work?

So we know an earthquake caused a dragon head to drop its ball into the frog's mouth below. But how, exactly, did that happen? It's not known for sure, but it's believed there was a **pendulum** inside the vase that moved according to the direction of the vibration. The swinging pendulum may have then moved rods or something else that prompted the release of the ball from the dragon head.

A Secret MILITARY AID

What did you imagine when you read this chapter's title? Did you picture some kind of weapon, perhaps something having to do with gunpowder or rockets? Well, prepare to be surprised. The secret military aid was the **wheelbarrow**. Yes, that's right—the wheelbarrow.

It's possible ancient Greeks used wheelbarrows as early as the 5th century BCE. However, that's not certain. What is certain is that the Chinese army began using wheelbarrows between 197 and 234 CE. Zhuge Liang is credited as the inventor.

The ancient Chinese army used wheelbarrows to carry food and weapons, to transport injured soldiers, and to set up obstacles in the path of attacking enemies. Wheelbarrows gave the Chinese army such an advantage over their enemies that they kept their knowledge of wheelbarrows a secret.

AN ANCIENT CHINESE WHEELBARROW IS DIFFERENT FROM THE KIND OF WHEELBARROW YOU'RE PROBABLY USED TO. IT HAD ONE LARGE WHEEL IN THE MIDDLE, WITH ROOM FOR GOODS ON EITHER SIDE OF THE WHEEL.

To keep wheelbarrows secret, ancient Chinese writings talked about them in code. For example, one report said, "Ko Yu built a wooden goat and rode away into the mountains on it."

Wheelbarrow Improvements

The Chinese found ways to improve the wheelbarrow so it could carry larger, heavier loads. One way was to have animals pull it while someone pushed it as usual. Another, more surprising, improvement was the addition of a sail to the wheelbarrow! Some sails were just pieces of cloth (see above). However, others were fully functioning smaller forms of the sails found on traditional Chinese sailboats and could be adjusted by the person pushing the wheelbarrow.

What TIME IS IT?

If you want to know the time, just check the nearest clock. Today, clocks—both digital and analog—are everywhere. But finding out the time in the ancient world wasn't so easy.

Some early timekeeping devices relied on sunlight, which wasn't around all the time. Others depended on water flowing out of or into a container at a constant pressure, which couldn't be guaranteed. Then, in 725 CE, a Chinese monk named Yi Xing invented the world's first mechanical clock. Inside the clock was a gold and bronze system of wheels, hooks, pins, shafts, locks, and rods. A stream of falling water set the clock in motion by turning a wheel. Every hour, a bell rang. And every 15 minutes, a drumbeat sounded.

THAT'S FASCINATING!

Yi Xing was actually working on a new, more accurate calendar and built an astronomical instrument to help with his work. That astronomical instrument just also happened to function as a clock.

ABOUT 250 YEARS AFTER YI XING INVENTED THE MECHANICAL CLOCK, CHINESE INVENTOR SU SONG CREATED HIS MECHANICAL CLOCK, SHOWN IN THE IMAGES HERE. IT WAS A HUGE DEVICE THAT TOOK UP SEVERAL STORIES IN A TOWER THAT WAS ABOUT 33 FEET (10 M) TALL!

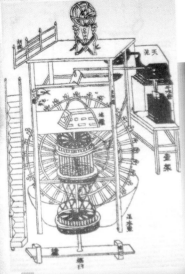

More About Su Song's Clock

Su Song's clock was considerably more elaborate than Yi Xing's. Yi Xing's clock had a bell and a drum. Su Song's clock had a large bell, small bell, drum, and **gong**. It also had four wooden **puppets** with movable arms to strike those instruments. In addition, it had 158 puppets to report the time. These puppets wore signs and colorful clothes in bright red, purple, and green.

The List

GOES ON

Although you've read about many ancient Chinese creations here, these are just some of their inventions. There are so many more, and many of them are part of modern daily life. Here are a few additional Chinese inventions.

The ancient Chinese invented acupuncture, the practice of sticking very fine needles through the skin at specific points to relieve pain and promote healing. It's very popular in the Western world today.

The ancient Chinese came up with the idea of planting crops in rows. This made it easier to irrigate and weed crops, and greatly increased crop yields.

What else? The ancient Chinese invented matches and porcelain—which is a very fine form of pottery. And more recently, the Chinese invented something you use every day—the toothbrush! Imagine your life without these and other Chinese inventions!

THAT'S FASCINATING!

The first toothbrushes were made of horse hairs attached to bone or bamboo handles.

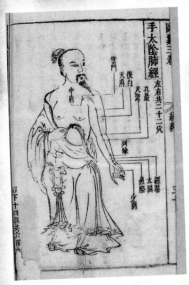

acupuncture chart

**ACUPUNCTURE
(AROUND 6000 BCE)**

**ROW-CROP FARMING
(6TH CENTURY BCE)**

**PORCELAIN
(25 TO 220 CE)**

**TOOTHBRUSH
(1498)**

**MATCHES
(577)**

**MOVABLE-TYPE
PRINTING
(11TH CENTURY)**

porcelain vessel

An Important Printing Invention

After paper, one of the most important Chinese inventions was printing with movable type. Earlier woodblock printing meant carving text into a block of wood for each page. Bi Sheng (990 to 1051) carved individual characters on separate pieces of clay, then hardened the clay in fire. These movable-type pieces could be assembled into text and glued to an iron plate to print a page, then separated and reassembled into new text for another page. This new process was much faster and less costly.

GLOSSARY

culture: the beliefs and ways of life of a group of people

dynasty: a family of rulers who have power over a country for a long period of time

gong: a large metal disc that makes a deep ringing sound when struck with a padded hammer

gunpowder: dry explosive matter that is used in guns and fireworks

hemp: a plant that is used to make thick ropes and clothes

magnetic field: the area around a magnet where its pull is felt. Earth has a magnetic field.

Middle Eastern: having to do with the area where southwestern Asia meets northeastern Africa

pasta: a food made from a mix of flour, water, and sometimes eggs that is formed into different shapes and usually must be boiled before eating

pendulum: a stick with a weight at the bottom that swings back and forth

puppet: a doll, often made of wood, with movable arms and legs

self-propelled: having the ability to move without aid

waterproof: able to keep water out

wheelbarrow: a small cart with two handles and one wheel

FOR MORE INFORMATION

BOOKS

Kovacs, Vic. *The Culture of the Qin and Han Dynasties of China*. New York, NY: PowerKids Press, 2016.

Morley, Jacqueline. *You Wouldn't Want to Work on the Great Wall of China! Defenses You'd Rather Not Build*. New York, NY: Franklin Watts, 2017.

Oachs, Emily Rose. *Ancient China*. Minneapolis, MN: Bellwether Media, 2020.

WEBSITES

Ancient China
www.ancientchina.co.uk/menu.html
Learn lots about ancient China on this interactive British Museum website.

Ancient China for Kids
www.ducksters.com/history/china/ancient_china.php
Find out more about ancient China on this site.

An Introduction to Ancient China
www.khanacademy.org/humanities/art-asia/imperial-china /neolithic-art-china/a/an-introduction-to-ancient-china
Discover more about ancient China and find a timeline here.

INDEX